Sybil Kane's unfinished cover

treetops1414@msn.com

Please visit us on the web @ insectwonderland.com

ISBN: 0-9777356-0-5

Library of Congress Cataloging-in-Publication Data

Application in process.

One World Press
1042 Willow Creek Road
Prescott, AZ 86301
800-250-8171

07 08 09 10 5 4 3

For retail orders please go to: insectwonderland.com
For bulk orders please email: treetops1414@msn.com

This book is a promise made and kept.
My work was completed while I slept.
As a butterfly's wings unfold,
At long last my insect poetry is told.

A is for ants
And aphids, too;
Who are the ants' cows,
Though they can't moo.

They're fed all summer
On Mama's pet plants,
Carefully tended
By farmer ants.

In winter time
They will be found
In an ant's barn,
Under the ground.

"Little aphids,
What can you do?
Why are the ants
So good to you?"

Sybil Kane

"Ants can do anything
Under the sun;
Anyway, anything
That can be done.

"Farmers, miners,
And carpenters, too—
'Most every job
That man can do."

"We give the ants milk
And that is why
We are so prized
And can live so high.

"We'd never change
Our masters—never!
These ants, they are
So very clever.

Sybil Kane

B is for bumblebee—
Her home's underground.
She rolls in the pollen
As she "bumbles" around.

B is for busy bee—
How she must work!
You'll never catch her
Trying to shirk.

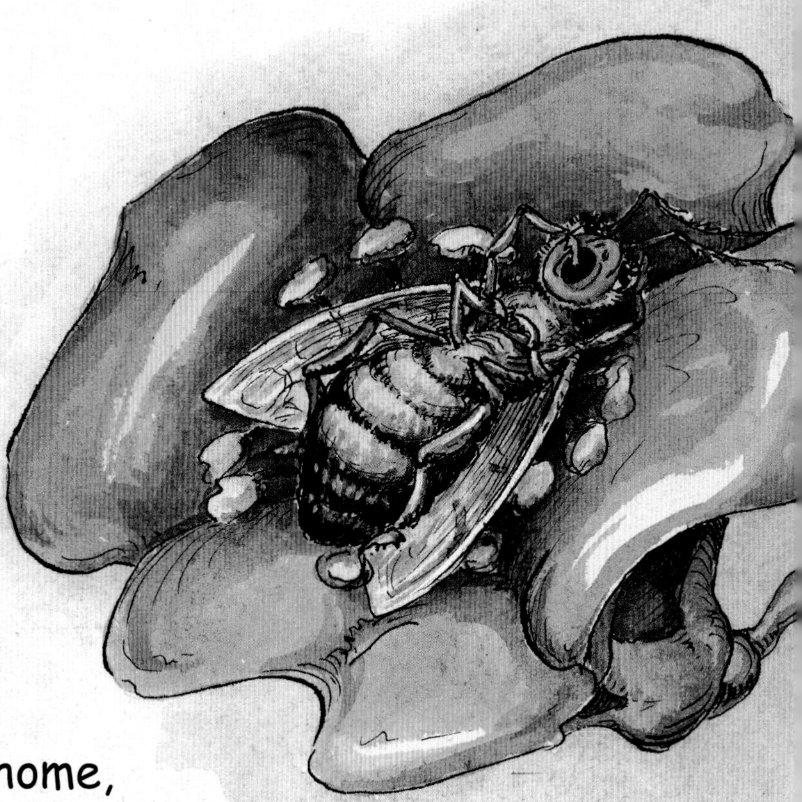

She lives in a hive
And builds in her home,
Out of six-sided cells,
A lovely wax comb.

In these cells she stores
Both pollen and honey
Brought from the flowers
On days long and sunny.

The ground bee's few cells
Lay in a flat row;
The honeybee's comb
Is so large
it might bow—

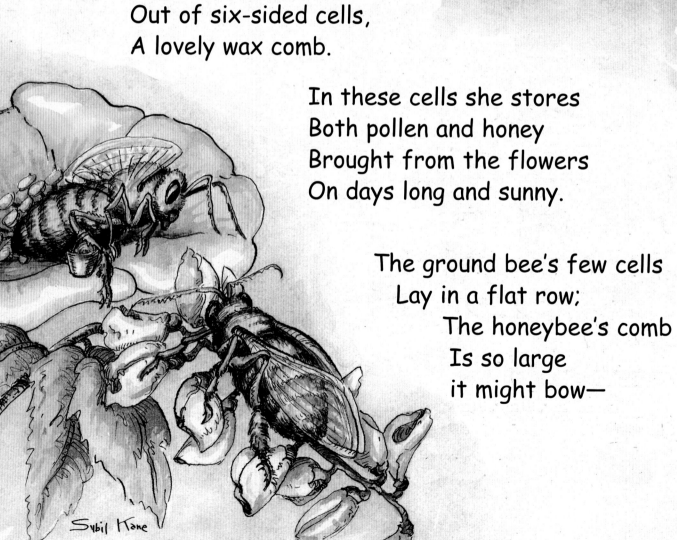

Or sag in the middle;
So, honeycomb wide
Is carefully built
Standing up on its side.

Bee architects know
It's far stronger this way.
Any "bee man" will show you
An upright bee tray.

Each beehive's a country,
Which has its own queen—
One fairyland that
Can really be seen.

C is for caterpillar,
Crawling on the ground,
Very soft and squishy,
Very fat and round!

Some are busy eating
Leaves high up in trees,
Holding very tightly
Whenever there's a
breeze.

They'll eat most plants,
Whether large or small.
Got to fill their tummies
Mighty full by fall.

C is for chrysalis,
A frail, cradle shell;
C is for silk cocoons,
Made so very well.

They are the little beds
For the caterpillar's night,
From which he will wake—
Quite changed to our sight.

A caterpillar's sleep
May last the winter
through;
What's happened by then
Seems too good to be true.

A butterfly or moth
Awakes in the spring;
God has made lovely
Each poor, crawling thing.

The creeping caterpillar
Just gobbled all his days;
The winged butterfly,
Among the flowers, plays.

D is for dragonfly,
But, oh me, oh my!
He isn't a dragon
And he isn't a fly!

When he is a baby,
He lives like a fish;
Darts about as gaily
As anyone could wish.

He uses jet propulsion,
For his peculiar whim
Is squirting water from his tail;
This is his way to swim.

He's rather like a crawfish,
In green or silver gray;
He doesn't pinch us with his claw,
Just tries to hide away.

He catches all the "wrigglers"
Living in the water,
That soon become mosquitoes
And things they hadn't ought'er.

If you go out wading
In the lily pool,
Please remember, killing friends
Is against the rule.

When at last he's grown,
He's changed in most things;
The best gift will be
His four stiff wings.

High over the cattails,
At great speed he'll race—
Half a mile in a minute
For this flying ace!

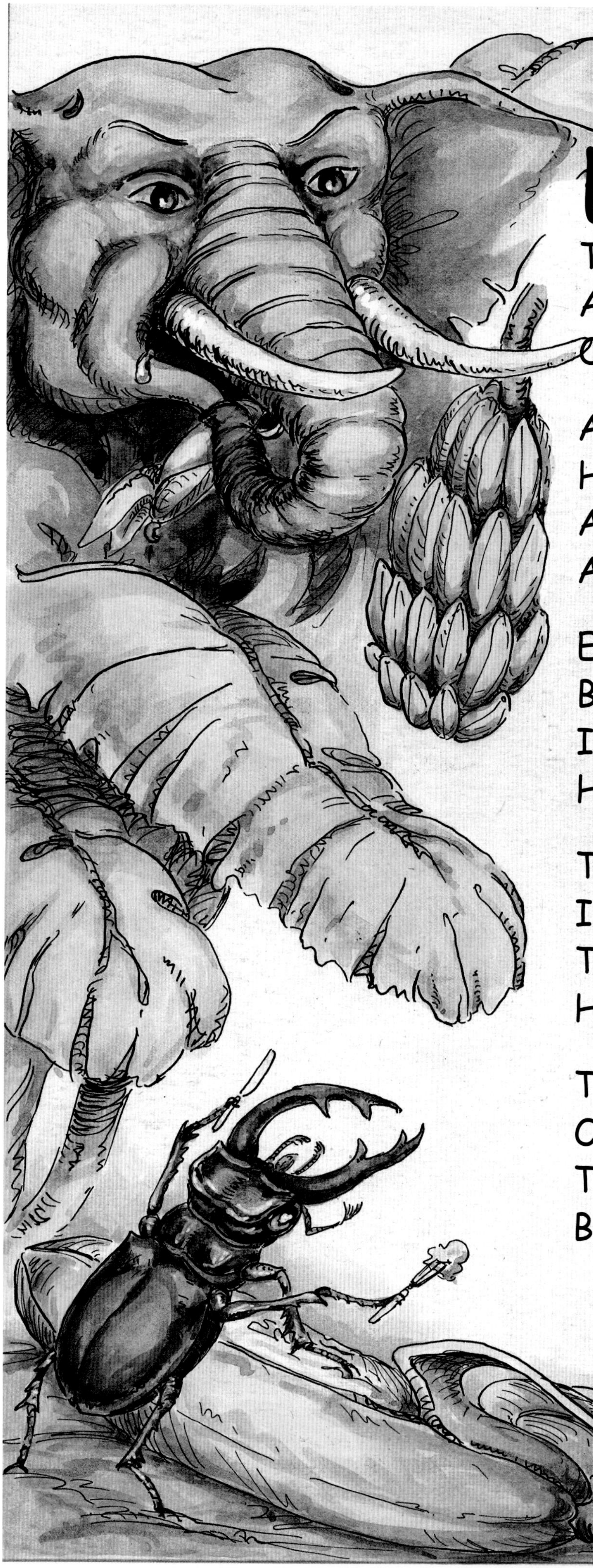

E stands for old elephant,
The very biggest beast;
And for the elephant beetle, too,
Of insects, not the least.

A handsome, shining fellow,
He's slow but very strong;
And, really, he is very big—
At least five inches long!

Elephants have shining tusks,
But this beetle's other name
Is staghorn, 'cause a stag
Has antlers just the same.

The elephants live
In countries very hot;
The elephant beetle loves
His South American lot.

To elephants the leaves
Of bananas are a treat;
The elephant beetle loves
Banana plants to eat.

When our beetle tries
His beetle wife to win,
He often finds another
Beetle way ahead of him.

Elegant elephant beetle,
Handsome brown and black;
A wife wants more than beauty,
Oh, alas and alack!

He must find in the armor
Of his foe a crack,
Or cleverly toss him over
Upon his sturdy back.

Leave him kicking, helpless
As a turtle in his shell,
Beaten and very angry
Yet quite unhurt and well.

Elegant elephant beetle
Now deserves his wife.
Let's hope this will end
All our beetle's strife.

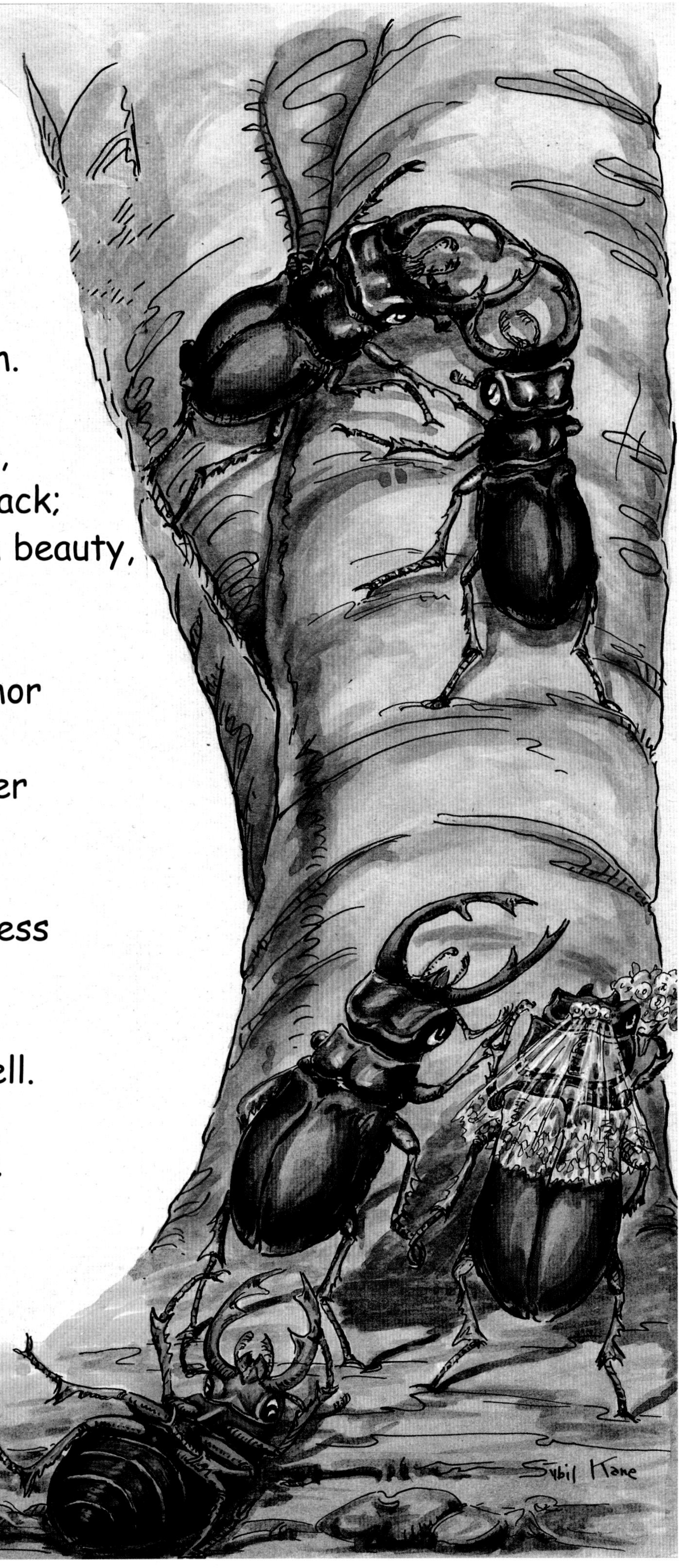

F — Why, that stands
For the horrible fly;
Quick, let's cover the food;
Don't let it come nigh!

It's not only the germs
That cling to its legs;
Each fly would lay hundreds
And hundreds of eggs.

They'd hatch into maggots,
Worms, very small,
Who'd eat until we
Had no food at all.

There are other flies,
Which are very much worse;
To horses and cows,
They're a terrible curse.

Horsefly and deerfly,
Ox-warble and gad,
Make very sore bites;
Or, what's really bad—

They lay eggs, which hatch
On the animals' hide,
Into larvae that bore
'Till they get inside.

These fly larvae make
Our friends very sick.
Let's spray barns and flies
With a fly poison—quick!

Don't think that all flies
Are fit just to kill,
For most of them work
As our friends—with a will.

Some are very bad pests;
But there's really no end
To the number who serve
As man's good friend.

Flies belong in the garden,
But not near our tables;
We also must keep them
Away from our stables.

G is for glowworms,
Glimmering in the gloom;
Bring them indoors
And they'll gleam
In your room.

Catch them very gently
And put them in a jar;
They will cast a light
As silvery as a star.

They are not worms
But larvae very small.
They haven't any wings
And cannot fly at all –

Until they are turned
Into fairy fireflies –
Those very lovely beetles
That flit across the skies.

Both make a lovely light
That flashes near their tails;
But it gives no heat
To cook their dinner snails.

They like to eat snails,
But in a manner most kind;
First, put them to sleep
So the snails don't mind.

It took our kind doctors
A great many years
To learn to stop pain
Like these little dears.

The poor snails are eaten
While still sound asleep;
But glowworms need food,
So we needn't weep.

Never chase the firefly.
As he flickers near and far;
He'll lead you to a swamp
Or strand you on a star.

The humble glowworms gleam
In the long, dark grass;
They light up the way
For you children to pass.

H is for horntail
In her suit of gold and black;
She does a lot of damage,
Oh, alas and alack!

She's rather like a wasp
In color and in shape,
But not with the good habits
We wish she would ape.

Of more use to the horntail
Than a chest full of jewels,
There is, in her tail,
A fine set of tools.

She carries two files
With which she can bore
Through a tree's bark,
Right down to the core.

And then a long tube
She can push deep inside;
Right there in the tree,
Her eggs she will hide.

They'll very soon hatch
Into little wood borers;
What they do to our trees,
Would give you the horrors!

They'll eat and they'll eat,
They'll gobble and swallow,
Make runways and caverns,
And leave our trees hollow—

Until they are ready
To make a snug nest
Of sawdust and silk
For their transforming rest.

When they awaken,
Like prisoners from jails,
They'll come to the surface—
Stripe-suited horntails.

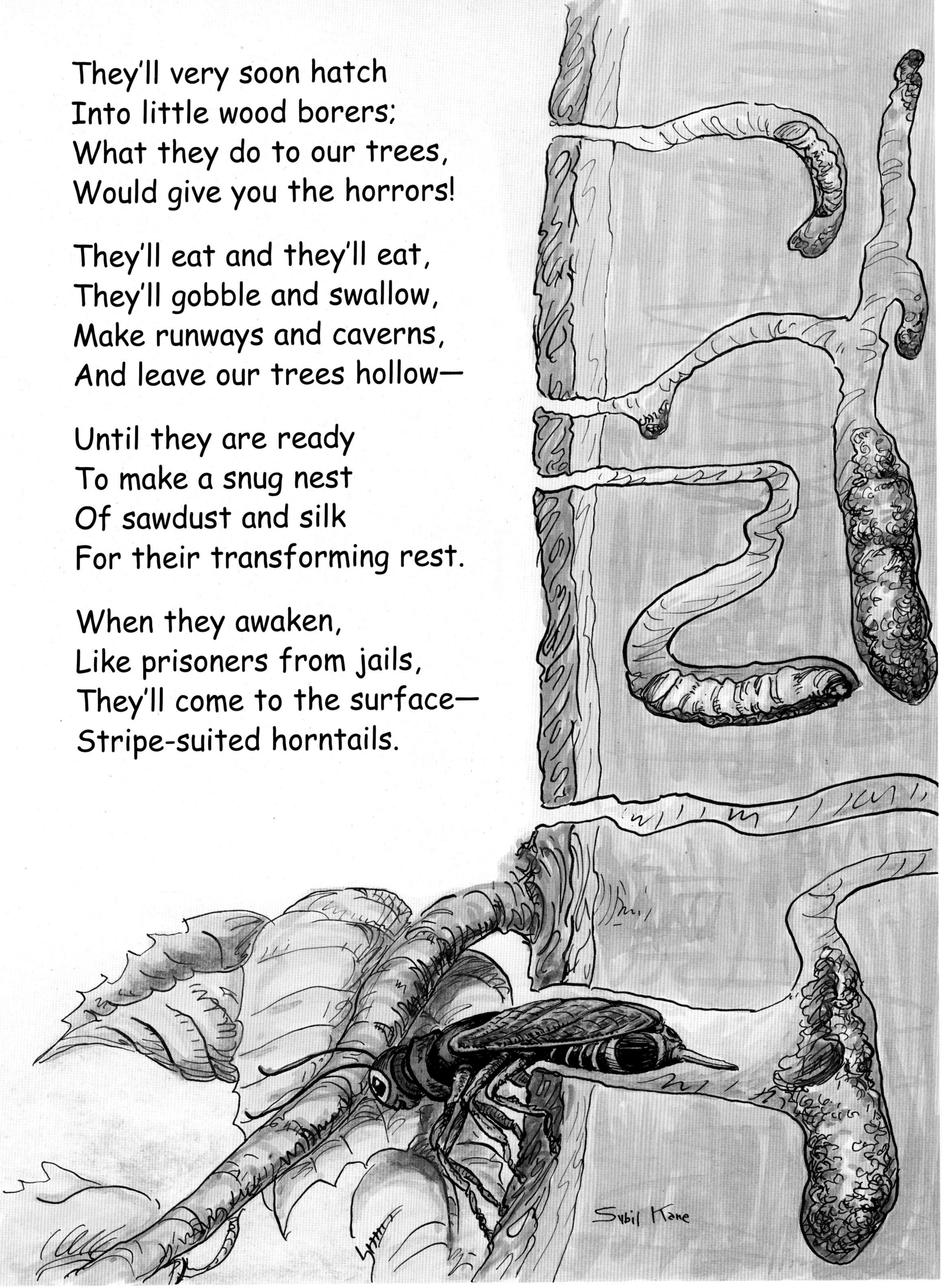

I is for inchworm,
Inching so slow.
"Do you measure everyone,
Everywhere you go?"

"Measuring worm, looper,
Inchworm, or spanner
Are names I am called
Because of my manner—

"Of looping along
And spanning an inch;
So measuring you,
For me, is a cinch.

"My colors are bright
And, though I'm not big,
They show when I move,
So I stand like a twig.

"Then birds pass me by;
And I am not seen
Any more than my brother,
Whose color is green.

"When I come down
To measure you all,
You might think I'd risk
A terrible fall.

"But, like a spider,
I'm able to spin
A long, silken thread
That's gossamer thin.

"I'll ride upon it
Until I can drop
On a boy or a girl
With a short 'kerplop!'

"I'll measure your trousers
Or your little bonnet,
And promise you new ones—
My word upon it!"

J is for a beetle
That came from Japan,
The Japanese beetle,
An enemy to man.

She's just a bit more
Than a quarter inch long.
She's fat and she's sturdy,
Her shell's hard and strong.

Her belly is striped
And her stiff wings, red;
Dull green is the color
Of her body and head.

She lays her eggs
In underground chutes;
The grubs that hatch out
Eat up our grass roots.

But killing our lawns
Is a very small part
Of the awful destruction
She makes a fine art.

The very worst damage
Is to apple and peach;
Not even the highest
Is out of her reach.

Two hundred or more
Crowd into one fruit,
Where each stuffs herself
Like a pig—the brute!

We spray and we spray
And still there are more—
Hundreds and thousands,
And millions, galore!

But in insect land
We've more than one friend
Who'll help us bring
This fiend to an end.

There are little wasps
And two kinds of flies
Whose larvae make sure
This horrid pest dies.

K is for the katydids,
Fine players all,
Who give us their music
Both summer and fall.

"Katy did, Katy didn't,
Katy didn't, Katy did"—
Comes the song from the grass
Where our dear friends are hid.

He really can't sing—
Just fiddles all day;
It truly is wonderful
That he can play—

Because his fiddle,
And fiddle bow, too,
Are in back of him;
That's where they grew!

One wing is the fiddle;
It's made to sing
When scraped by the edge
Of the fiddle bow wing.

The husband can play,
But not little Kate,
Who must listen all day
To the song of her mate.

She hears his gay music,
With ears on her legs,
As she tends to the business
Of laying her eggs.

The crickets, too, play
A gay little song,
And the grasshoppers, too,
When their "horns" are long.

Ask weather cricket—
"How hot is the day?"
Then count for a minute
The chirps he will play.

Divide them by four
And then forty add;
The temperature now
You will know, my lad.

L is for the ladybird,

A beetle you should know
From the verses made about her
In Europe years ago.

For killing pests like aphids,
There were not any sprays;
The lady and her babies
Cleaned the hop vines in those days.

When the hop vines were burned,
She was told to go home
To rescue all her babies
Who had no wings to roam.

She's called Our Lady's beetle;
Was, by the Virgin, blest
For ridding farm and garden
Of many a horrid pest.

Scale, she eats, and aphids,
That are tended by ants,
Pests that suck the juices
And so kill our plants.

Her grubs change their jackets
Three times as they grow
And finally turn into
The beetle you know.

Quite late in the fall,
When days colder grow,
The beetles all fly
To hills near the snow.

There, many thousands
Are sure to be found,
Ready for winter,
Asleep on the ground.

Men gather them up
And store them on ice
Until they are needed
To eat the plant lice.

Then, they'll be shipped
South, east, or west—
Where they will wake up
And go after each pest!

M is for mosquitoes,
Singing in the dark;
Keeping us awake
Is their greatest lark.

Turn on the light
Or, quick, get under cover!
Hordes of hungry "skeeters"
Over your bed do hover.

Mosquitoes do not bite you,
In spite of what you think;
They need our blood to live,
So they must take a drink.

Not with a mouth like ours,
But with a needle sharp and hollow,
The female jabs in deep
To take a great big swallow.

To treat us this way
Seems just awfully wrong;
The "skeeter" up north
Will buzz her warning song.

Breathing tubes in the bodies
Of each "skeeter," fly, and bee
Are blown like tiny flutes
To warn both you and me.

Our "skeeter" has a cousin
Who's called anopheles;
She stands so very oddly—
As though down on her knees.

She's evil and a coward,
No warning song she sings;
Malaria and fever,
With her bite she brings.

She breeds in the South,
In water where it's hot;
To kill her, men have sprayed
The swamplands quite a lot.

She hatches out from "wrigglers,"
In stagnant bits of water;
Our part's to clean up tin cans
And such things, as we "ought'er."

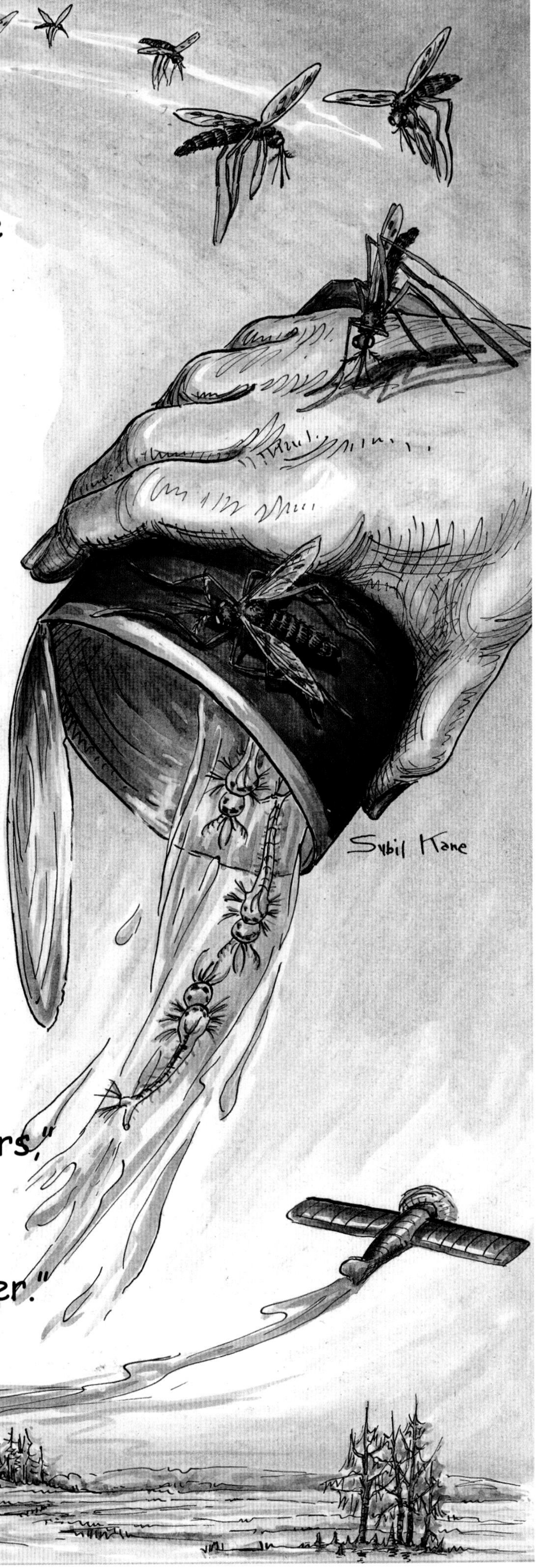

N is for nimble fly
Who helps us very much
To kill the caterpillars,
Beetles, worms, and such.

I've told you before,
We have many a friend
Among all the flies
Who help us no end.

Tachina and nimble flies
Are the best of all these;
They fly among the vegetables
And flowers—like the bees.

On army worm and locust
And caterpillar skin,
They lay the eggs, which hatch
Into larvae that bore in.

These larvae are worms,
Very hungry, very small.
They bore through each pest's skin;
But that isn't all—

Into the snug cocoon,
It makes for its nest,
The larvae are carried
By each sleepy pest.

The larvae will eat
The pests while they sleep;
The pests won't hatch out
To make farmers weep.

From each pest's cocoon,
When springtime is near,
Hatched from the larvae,
New flies will appear.

Great armies of locusts
Would destroy all our grain;
The worms and the grubs
Give all farmers a pain.

There also are beetles,
Like the bad Japanese;
You see how the good flies
Can help us kill these.

O is for the oak-boring grub,
Who spends most of her life in the oak;
Like all the long-horned beetle's grubs,
The damage she does is no joke.

In oak tree bark her mother hid
An egg you will hardly see;
Although it's a Capricorn beetle's egg,
It will hatch an oak-borer wee.

The mother is dead, now the tiny grub
Has no one there beside her;
She has no eyes, she has no ears—
Only instinct to guide her.

A set of tools in the top of her head
Is made for working wood;
She'll tunnel inside the oak tree for her
house and for her food.

Three years she'll live
in the darkness deep;
Does she dream a beetle
she'll be,
After she's had her
transforming sleep,
And live outside the tree?

What tells her when the time has come
To tunnel up to air;
Then hurry down and build herself,
Of silk and chips, a lair?

For safety, she will make three doors
Behind which she'll be found
Deep in a sleep in which she must
Never turn around.

She'll be a beetle, when she wakes,
Who can't turn in the tree.
Her new shell is too hard for that,
As you can plainly see.

Instinct taught her the tunnel to make
And keep the door at her face;
The years of darkness now are passed;
To the surface she can race!

All the long-horned beetle grubs
Destroy many thousands of trees—
Locust, elder, apple, and oak
Are just a few of these.

P is for praying mantis,
An insect very odd;
She's named that for she
often stands
As though in prayer to God.

Her body is three inches long,
Wood-brown and strange in shape;
Her wings, a lovely shade of green,
Just like the leaves they ape.

Her four back legs are twig-like;
But her front "praying" claws
Aren't as innocent as they seem—
They're pronged and look like saws.

Yet they could not defend her
Against her big bird foes,
But leaf-like wings and stick-like legs
Are most protective clothes.

Because her foes can't see her,
She's safe from all but man.
Let's you and I not hurt her;
She'll help us when she can.

She does not harm the trees
Or eat the leaves and bark;
She makes a fine policeman
In any woods or park.

She eats a hundred little bugs;
Then up comes something
fierce—
An enemy that she must fight,
Whose armor she must pierce.

She'll rear up on her back legs,
Her front claws in the air;
She holds them as though she were
About to say a prayer.

Her fierce foe soon will charge her
And she must ready be
To pounce on him and grab him
And battle manfully.

Those claws of hers will hold him
While she bites him quite apart.
Her foes were fooled by "praying"
arms;
Our lady has no heart!

Q is for queen bee;
Of her hive, the mother.
She lays all the eggs
But has no more bother.

Just once in her life
Does she go out to fly,
And that's for her wedding
Up high in the sky.

The husband bee dies
But the queen comes home
To lay thousands of eggs
And never more roam.

She's fed royal jelly,
And kept from all harm,
Even carefully fanned
When the weather's too warm.

Her sisters, the workers,
Make the hive home,
Placing pollen and honey
Inside the wax comb

They tend her bee babies;
Three days all are fed
On royal milk jelly
From kind nursie's head.

The baby queens eat
Royal jelly alone.
The others eat pollen
And sweets from the comb.

After three more days,
Wee cocoons they make,
Take a twelve-day sleep—
And, as bees, will awake.

They'll clean up the hive
Queen and her babies tend,
Turn nectar to honey,
Their worn homes they'll mend.

When strong enough now,
For food they must fly.
For their queen and her babies,
They'll live and they'll die.

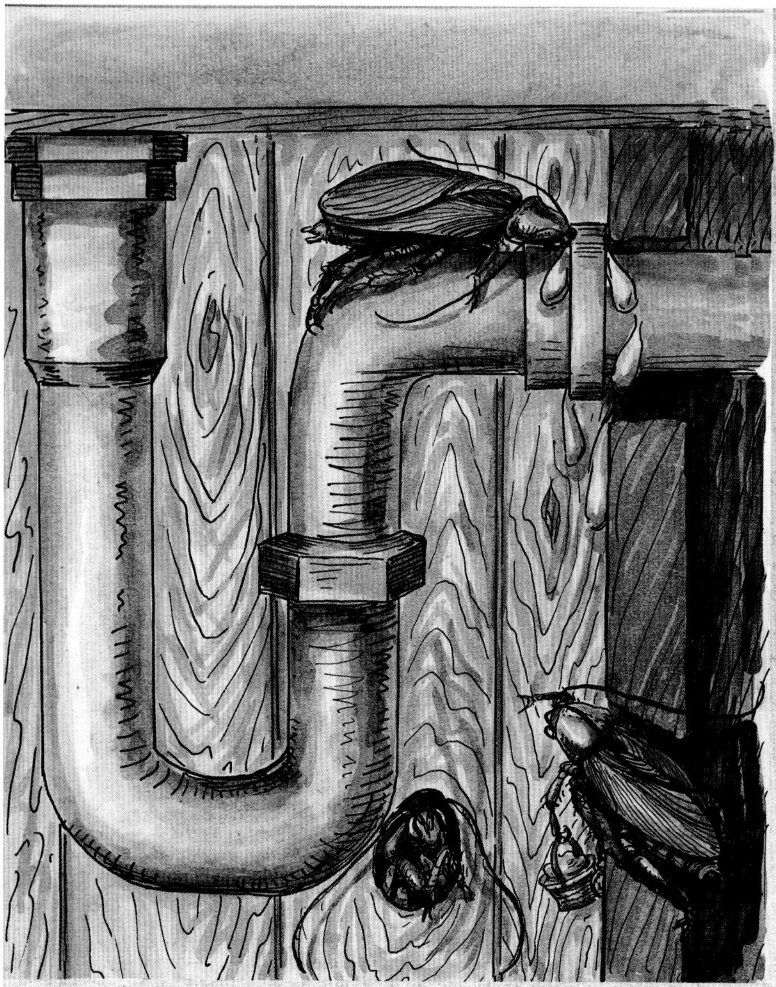

Their family was on earth
Millions of years ago,
Living in fern forests;
The coal beds tell us so.

Now the ferns have turned
To coal, deep underground,
But roaches of today
Are like the fossils found.

In mankind's earliest days,
The roaches ate their food;
They still search for it today,
Because they find it so good.

R is for the roaches—
There, behind the sink!
They hide in kitchen crevices
Before you even wink.

Put out cockroach poison,
Like old "gator hives;"
But it won't surprise me
If sturdy roach survives.

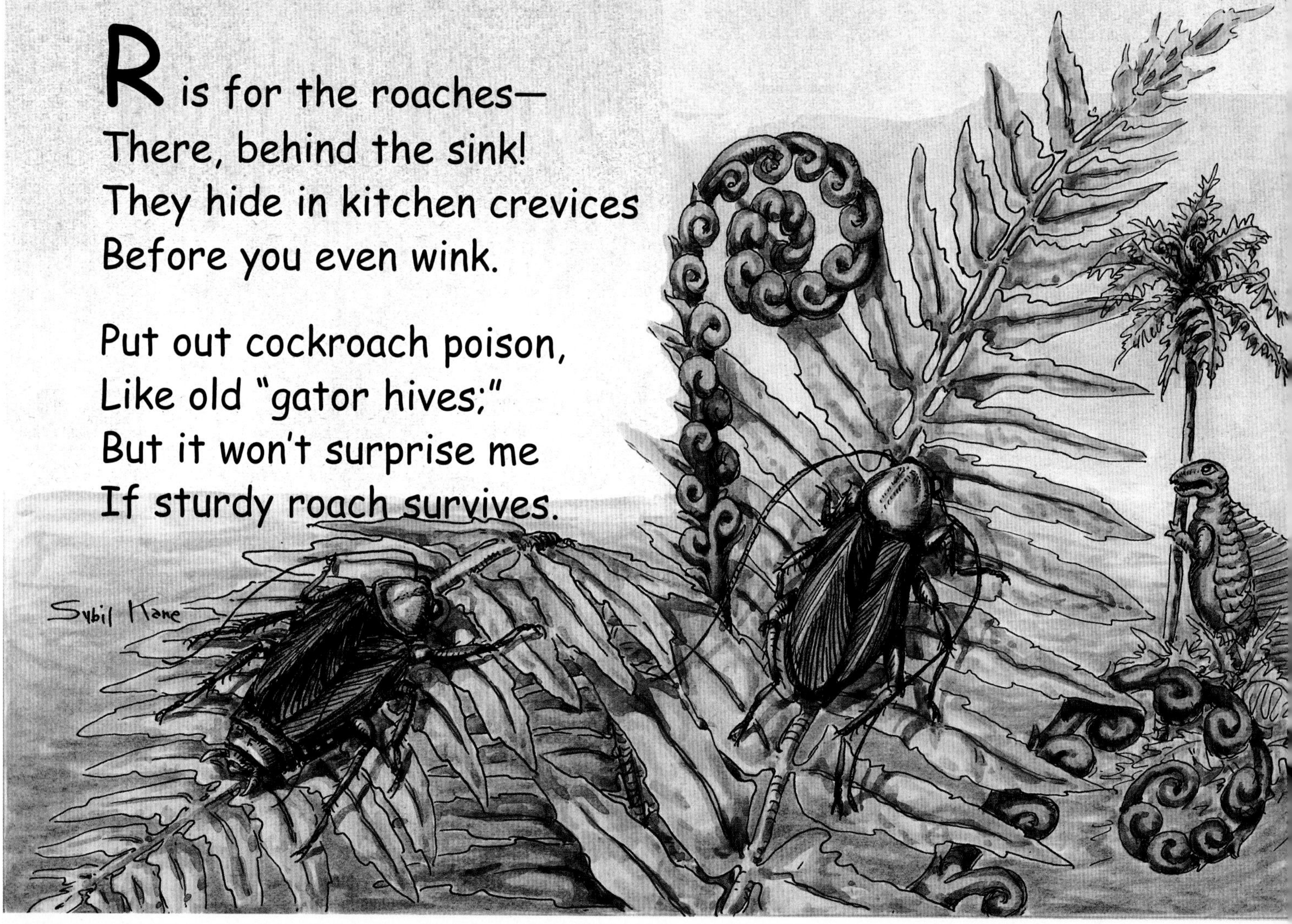

Long ago in Egypt,
When pyramids were the style,
They found food in boats,
Afloat upon the Nile.

They've sailed around the world,
Mostly in ships of wood.
They've settled in every land
To eat up human food.

They also have a love
For human fingernails;
Ask any sailor who
A tropic freighter sails.

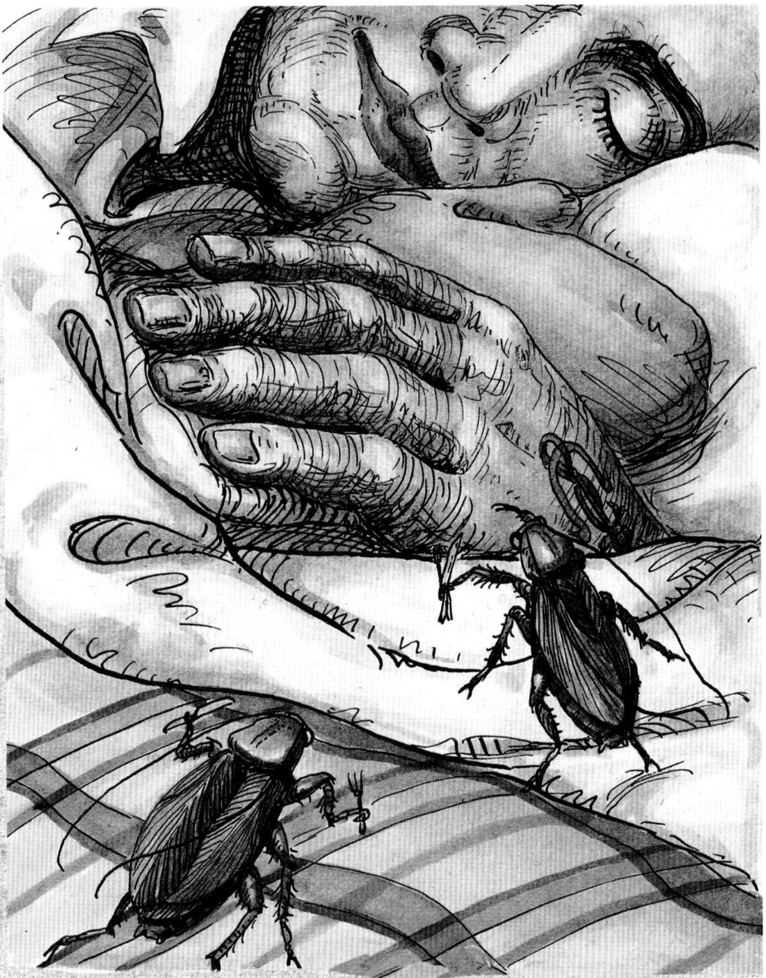

On his hands and feet,
He must nightly keep
Pairs of socks and mittens,
For safety when asleep.

Let us rid our kitchens
Of this horrid pest;
He's dirty and a thief
Even when at rest.

S is for the spider;

I'll put her in this book.
Although she's not an insect,
Let's take a careful look.

All insects have three parts—
Belly, thorax, and head;
Some can spin while larvae,
From their mouths, comes the thread.

But spiders have no necks,
Thorax and head are one;
From tubes behind their bellies,
Their silken webs are spun.

Three pairs of legs have insects,
But four pairs has the spider.
She's a cousin of the crabs;
Put a sand crab beside her.

She has some other cousins—
The scorpions, ticks, and mites.
They are unpleasant fellows,
With nasty stings or bites.

The venomous black widow is
A villain from a "thriller."
Before she can poison you,
Brave spider wasp will kill her.

The others will not harm you
But do a lot of good.
They only use their poison
To kill their insect food.

Crab spiders run sideways,
Just like the real crab!
Wolf spiders can jump—
Catch food with a grab.

The rest of them spin,
With patience and skill,
The most wonderful traps
For the insects they kill.

In gardens, on grass,
All over the place—
Oh, children, do study
Their patterns of lace!

T is for the termites small.
In spite of what folks say,
They're not "white ants," or any ants,
And differ in many a way.

They tunnel through our house walls
To make each termite city.
They're clever like the ants and bees;
And, really, it's a pity—

That humans must destroy them
For hollowing out the wood,
To make their termite colonies,
And also for their food.

Each year the winged termites fly,
The young kings choose their wives,
Bite off their wings, and settle down
For all the rest of their lives.

Each pair will chisel out a room,
Lay eggs and babies feed,
Enlarge their homes as the babies grow,
And see to all they need.

In a year or so the parents can
Become the king and queen,
Be cared for by blind workers,
Fed, and kept so clean!

The workers never leave their homes
And have no need for sight,
Are guarded by big-jawed soldiers
Who protect them day and night.

There are prince and princess nymphs
Who have no wings to fly
But can replace the king or queen
If either one should die.

Termites can change their babies
Into just whatever they please—
Soldiers, workers, kings and queens,
Or nymphs, with the greatest of ease.

These termites are so clever,
Admire them we do.
But they destroy our houses,
So we must kill them, too.

U is for the underwings,
Bright-winged moths are they,
Who can escape their enemies
By hiding their color away.

Folding their wings down tightly
When they want to rest,
Is the method that all moths
Seem to find best.

When God, who created beauty,
Gave to these moths their share,
He only made their underwings
A brightly colored pair.

The upper wings are patterned,
So much like a tree's bark,
That folding them, in hide or seek,
Can almost be a lark.

They really are in danger
When in the air they soar.
Now birds that choose bright colors
Can't see them anymore.

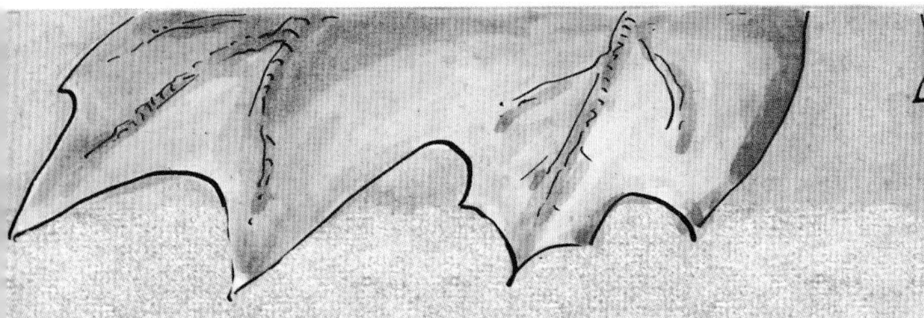

All moths live to lay their eggs—
Some do it right away;
Those are born without a mouth
And die on their first day.

Others can live four months or more,
Feeding on nectar sips,
Taken through their siphon tongues
From the flowers' lips.

A hundred kinds of moths have grubs
That are a horrid curse.
There are nine thousand harmless ones,
So it could be much worse.

Moth antennae have no knobs—
They're pointed or they're feathered.
Moth cocoons are made with silk
And sometimes leaves all weathered.

A few moths are brightly colored
And fly in the afternoon;
But most are light or colored white
And dance beneath the moon.

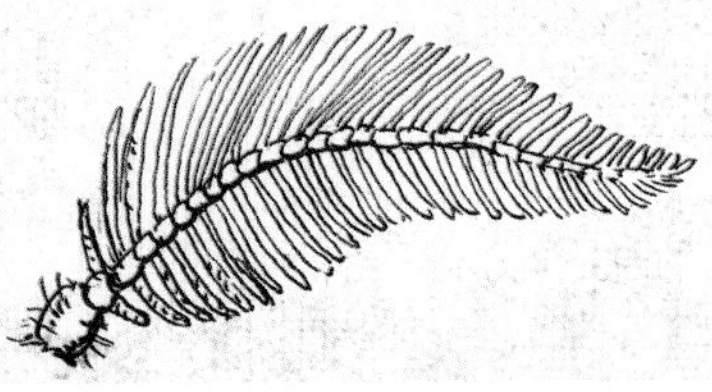

V is for Vanessa,
Butterflies rather rare,
Who hibernate each winter
Like a woodchuck or a bear.

Most butterflies sleep in their
chrysalids,
While others sleep as eggs.
Some caterpillars sleep
With silk blankets 'round their legs.

One butterfly, the monarch,
Can to the southland fly.
Only our dear Vanessas sleep,
As butterflies, nearby.

Hatching from their chrysalids,
When autumn leaves are gay,
They must find a place to pass
The winter, cold and gray.

At times near the end of winter,
When the sun is hot and bright,
They will wake and venture out
On a short but lonely flight.

Then they must return and sleep
Until the early spring,
When strange and lovely messages
Their sense of smell will bring.

Like radio waves to the airplanes,
This delicate scent is their guide.
They'll find their mates at the
    end of it,
Though the sky is very wide.

Each butterfly's antenna
Has a knob upon its tip;
Her body's slim; her tongue's a
    tube,
For nectar she will sip.

The mourning cloaks and admirals,
Commas, and angle wings
Make up the Vanessa family;
All rest with folded wings—

As do all other butterflies
Who can, in no other way,
Hide the brightness of their wings,
Since they always fly by day.

W brings you the water bugs—
You should know seven at least.
Their antics are amusing, but
One is a murderous beast.

He's called the giant water bug
And colored a brownish-yellow.
The great black diving beetle
Is an even fiercer fellow.

Both breathe from pores beneath their
wings—
Under water they must have air.
They carry a bubble down with them,
Held in their long, back hair.

They kill most bugs and lots of fish,
Can fly from pool to pool;
Their bite is quite a painful thing;
They're nothing with which to fool.

The water scorpion pokes his tail
Out of the water for air.
It has no sting like the real scorpion's—
The scorpion name isn't fair.

He sucks the juices from other bugs
With a sucker that could pierce
Even the skin of humans;
But you'll never find him fierce.

Next comes the water boatman—
He has back legs shaped like oars.
He has wings, too, and many times,
Over his home pond he soars.

Back swimmer swims upon his back—
It's easy for him to float,
Because he carries his breathing air
Around him like a silver coat.

The water striders are great fun—
With four legs spread out wide,
They'll slide upon the surface in
A skater's graceful glide.

The beetles that in circles dash
Are comical whirligigs;
See how the surface ripples as
They dance their merry jigs!

X is for xyllo-
Copidae,
Which only means
The carpenter bee.

A carpenter bee!
How strange, you say.
Well, it is because
Of her wood-working way.

Her home is not
In a hive so big,
But a burrow deep,
Which she must dig.

She makes this hole
One-half-inch wide,
In lumber, dry,
Or a dead tree's side.

She bores half an inch
Against the grain,
Which is deep enough
To keep out the rain.

Then, sharply she turns,
With the grain she'll go—
Some twelve more inches,
Sometimes, twenty or so.

If a man with his teeth
A tunnel could dig,
It would be ninety feet long
To be as big.

She makes a snug home
And stores her food,
And raises her babies
Deep down in the wood.

She's as clumsy and big
As the bumble bee;
Can enter the flowers
No better than he.

But teeth that can chisel
Deep in the wood
Can open a blossom
For nectar, so good.

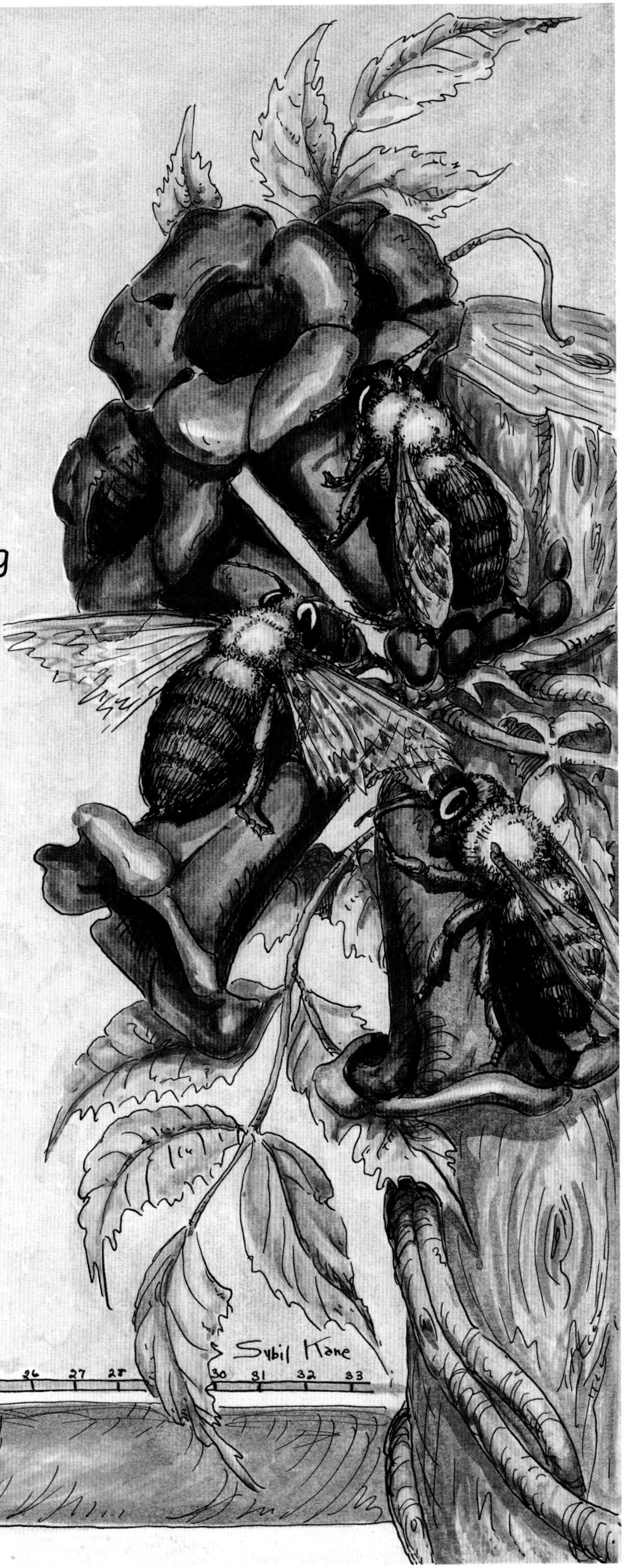

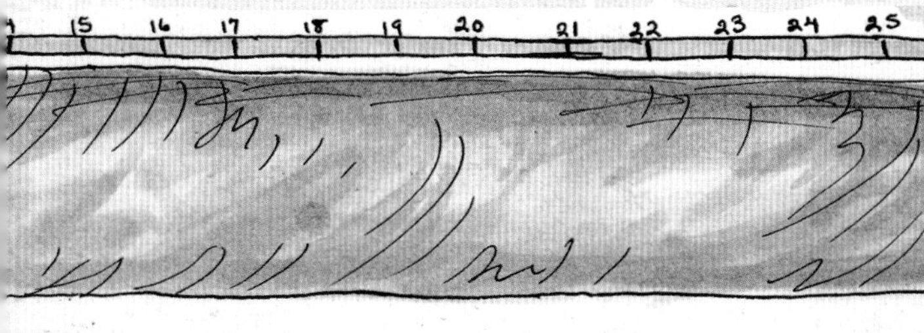

Y - for the yellow jacket stands;
A "social" wasp is she.
She's called that for, like bees' and ants',
Her home is a colony.

Paper wasps, hornets, and "yellow jacks"
Are the social wasps best known;
Each chews up flakes of old, dry wood
To make her paper home.

Shaped like the waxen comb of bees
Are cells for their babies made;
But they're of paper and upside down.
And, when an egg is laid—

It's firmly glued to the sides of its cell
So that it won't fall out.
All the time it grows as a baby grub,
It must cling with muscles stout.

These wasps kill many harmful bugs.
They chew them into paste;
They feed this to their babies, plus
Fruit juices—just a taste.

Paper wasps' cells hang upside down,
Exposed to cold and storm;
The hornets' nests, strong paper walls,
Make home quite safe and warm.

"Yellow jack's" nest has thinner walls,
Because it is hidden away
In an animal's burrow or hollow log.
It's safer there—happen what may!

The social wasps invented paper,
But this is also true:
Man's first dishes were made by copying
Mud daubers and potter wasps, too.

Caterpillars and spiders are stored, asleep,
In these nests they make of clay—
Food for wasp babies, when they hatch out,
Who are fed in no other way.

In spite of the wonderful things they know,
Each winter most wasps must die.
Just a few lonely queens start over each spring;
The reason? I can't tell you why.

Z is for zygoptera.
They'd have a lot more fame
If they were just called damselflies,
Which is their other name.

The dragonflies' first cousins,
They differ in many ways;
Though both first live under water
And in the sky end their days.

Zygoptera's nymphs aren't jet-pro-
pelled—
Their bodies are slim and frail;
They have to swim and also breathe
With feathery gills in the tail.

On plants beneath the water,
They fish, as their cousins do;
Catch wrigglers with their funny claws
When they swim into view.

When they turn into damselflies,
They're able to fold their wings.
You'll see them clinging to branches—
Like rubies strung in rings.

They often fly together—
Eight-winged flying ships;
The husband guides his wife
On her egg-laying trips.

They must plunge under water.
It's there her eggs she'll lay,
In slits both make in plant stems,
To tuck each egg away.

It's hard to climb through water
To the lovely world above;
The husband climbs his very best;
The wife gives a mighty shove!

Once he is out of the water,
He flies with all his might,
And pulls his wife right after him,
Back to the world of flight.

Z is the end of the alphabet
But I hope, from this day,
You'll watch the insect wonderland
When you go out to play.

Firefly Fairy

Marguerite Young